Publicity stunts of all kinds abounded whenever a new motorcycle came on to the market. Here a 398 cc 1920 ABC is put through its paces with five people aboard, to prove it was more than capable of carrying a heavy load. Unusually, the photograph was not taken whilst the machine was ascending a steep hill, which would have been more convincing. The engine was a horizontally-opposed overhead-valve twin.

VINTAGE MOTORCYCLES

Jeff Clew

Shire Publications Ltd

CONTENTS

Published in 1995 by Shire Publications Ltd, Cromwell House, Church Street, Princes Risborough, Buckinghamshire HP27 9AA, UK. Copyright © 1995 by Jeff Clew. First edition 1995. Shire Album 314. ISBN 0 7478 0277 7. Jeff Clew is hereby identified as the author of this work in accordance with Section 77 of the Copyright, Designs and Patents Act 1988.

Printed in Great Britain by CIT Printing Services, Press Buildings, Merlins Bridge, Haverfordwest, Dyfed SA61 1XF.

British Library Cataloguing in Publication Data: Clew, Jeff. Vintage Motorcycles. – (Shire Albums; No. 314). I. Title II. Series. 629.227509. ISBN 0-7478-0277-7.

Editorial Consultant: Michael E. Ware, Curator of the National Motor Museum, Beaulieu, Hampshire.

ACKNOWLEDGEMENTS
My grateful thanks to Geoff Morris, a former Pioneer Machine Registrar for the Sunbeam Motor Cycle Club, who kindly read my manuscript and made a number of useful suggestions about its content. My thanks also to the Photographic Library of the National Motor Museum at Beaulieu, who allowed me to select and use photographs from their archive. All photographs are reproduced by kind permission of the National Motor Museum.

Cover: A 1921 247 cc Cedos two-stroke made by Hanwell & Sons Components Ltd of Northampton, photographed at Hamworthy, Dorset. The model was named after the brothers who designed and made it, Cedric and Oscar Hanwell, and only about five examples survive. It was restored by Ken Sackley, of Hamworthy, Dorset.

Motorcycle racing on public roads was banned in 1925 after an accident on Kop Hill, near Princes Risborough, Buckinghamshire, when some spectators were injured. This photograph of Freddie Dixon ascending the hill with a sidecar outfit shows how spectators lined the road, with no protective barriers or any attempt at crowd control.

Clyno and Royal Enfield vee-twins were ordered for use by the forces during the First World War, either for transporting an officer (as seen here) or with a sidecar on which a machine gun was mounted. Following behind in this photograph are despatch riders riding single-cylinder Model H Triumphs. The War Office had recognised the importance of mechanised transport in modern warfare.

WAR AND THE AFTERMATH

When war broke out in 1914, the internal combustion engine had become sufficiently well-established to play a leading role. The aeroplane and the tank came into their own during this conflict and, as telecommunications were primitive, the despatch rider and his motorcycle became indispensable. Commandeering civilian machines had proved unsatisfactory, so the War Office contracted a number of manufacturers to supply machines for use by the armed forces. The largest orders for solo machines went to Douglas and Triumph, although P & M supplied some to the Royal Flying Corps. Royal Enfield, Clyno and Scott supplied larger capacity models, with a sidecar that had a mounting for a machine gun. By the end of the war Douglas had supplied 25,000 and Triumph 30,000 to the War Office. Strangely, the production of motorcycles for the civilian market was not terminated until 15th February 1917.

The countershaft gearbox had superseded the epicyclic hub gear and, because it had a kickstarter, pedals were no longer needed. It was more robust, gave a wider range of gear ratios and was easier to dismantle and repair. Douglas had fitted a countershaft gearbox since 1910, and Triumph followed suit in early 1915, when they introduced their Model H.

The war ended the supply of German-made Bosch magnetos, fitted by many manufacturers. As a result, Dixie and Splitdorf magnetos had to be imported from the United States, until British electrical manufacturers could supply their own in sufficient quantity.

When the production of machines for the civilian market resumed in 1919, there were few new designs. Most manufacturers had little opportunity to devote time to them when they were working flat out for the war effort, although a few had experimented with spring frames. The majority carried on more or less where they had left off before the war, with only minor detail changes to their pre-war models. New models were difficult to obtain, and expensive, so the more unscrupulous dealers turned this situation to their advantage, charging a premium to move a customer's name higher

The Model H Triumph was virtually a 1914 550 cc model fitted with a modified frame to carry a three-speed Sturmey Archer countershaft gearbox. It was a great improvement over the earlier epicyclic hub gear, also made by Sturmey Archer. The Model H, though announced during late 1914, was not produced until February 1915. Note the capacious box mounted on the rear carrier and also the circular leather box that carried a spare final drive belt.

up the waiting list. Many, especially the more impecunious, opted for a former War Department machine, of which there were a great many available.

Gradually the motorcycle market returned to a state of normality, and prices stabilised. Entirely new models began to appear at last, amongst them flat twins made by manufacturers who had not considered this type of engine layout before. They included Humber, Raleigh and, in the United States, Harley-Davidson. The most promising of them all seemed to be the 398 cc ABC, made in the Sopwith factory, where fighter aircraft had been produced during the war. Unlike a Douglas, it had an overhead-valve engine, mounted transversely in a spring frame. Production and reliability problems limited its popularity and by 1922 it had gone.

Tom Sopwith enticed Walter Moore from Douglas to help get the post-war ABC into production in his factory at Kingston-upon-Thames, where Sopwith aircraft had been built during the war. The initials of this machine were a contraction of All British (Engine) Company. Designed by Granville Bradshaw, the 398 cc post-war model had its engine mounted transversely, not longitudinally, as the earlier 500 cc model.

This Kingsbury scooter advertisement is self-explanatory. With no provision for seating, it was obvious the scooter had been designed with journeys of only short duration in mind. Being of small diameter, the wheels would not have aided stability on the unmade roads of that era.

Scooters also appeared at this time, seen more as a novelty than as a serious means of transport for short journeys. Most had no seat, the rider having to stand throughout the journey. This, the lack of a gearbox and the small-diameter wheels made them unstable. They were soon little more than a memory.

At the beginning of 1921, following the Roads Act of 1920, every motor vehicle had to be registered, and a registration book issued in which its details were recorded. It had to be produced when purchasing a now obligatory road-tax disc, to be displayed on the vehicle. Motorcycles weighing up to 200 pounds (90 kg) were taxed at £1 10s per annum, and

ABC produced the 125 cc Skootamota in Sopwith's Walton-on-Thames factory. With its rear-mounted four-stroke engine and sprung saddle, it was a much more practical scooter than most of its contemporaries. Skootamotas were road-tested by riding them to Brighton and back, a quite severe (and time-consuming) test for such a small-capacity machine.

Lightweight two-strokes were coming very much into their own at this time, many of them powered by a Villiers engine. A notable exception was the Levis, fitted with the company's own engine. Designed by Bob Newey and made by the Butterfield brothers in Stechford, Birmingham, the Levis was a leading name in the industry. The rider is Geoff Davison, who won the 1922 Lightweight on one of these 250 cc two-strokes.

over that weight £3. Sidecar outfits and three-wheelers weighing up to 8½ cwt (432 kg) were charged an extra £1. Some owners displayed a Guinness beer-bottle label, which was the same colour as a tax disc, and cheaper! The nationwide speed limit of 20 mph (32 km/h) remained unchanged, a somewhat unrealistic speed which few bothered to observe. Speed traps helped to enforce it, a tactic that did not earn the police much respect.

The small-capacity two-stroke was now enjoying a spell of popularity, with a large number of different makes on the market. The majority had a Villiers engine and either a countershaft or a two-speed gearbox, but there were other proprietary engine manufacturers who could offer an alternative. A belt was still the most favoured final drive. Most two-strokes ran on petroil, a predetermined amount of oil being added to the petrol, to lubricate the engine's internals.

Dynamo lighting systems had been in use since 1914, but the majority of motorcyclists still relied on acetylene lighting. By dripping water on to calcium carbide, acetylene gas was generated and passed by a rubber tube to the headlamp and the tail lamp, where it would burn quite brightly.

Norton Motors of Bracebridge Street, Birmingham, had a very long association with racing, having won the Twin Cylinder Class of the first TT in 1907. They sold two specially tuned versions of their 490 cc single-cylinder side-valve model, the BR and the BRS, which were supplied with a certificated performance. D. R. O'Donovan, a record-breaker at Brooklands, tested the engines there in this 'slave' frame, to provide the figures for each certificate.

BMW originally made aircraft engines but after the First World War they began making a horizontally opposed twin-cylinder engine along the lines of the Douglas. In 1923 they produced their first complete motorcycle, the R32, with shaft final drive and leaf-spring front suspension. It set the pattern for what was to follow for many years to come.

It was a simple system to use, provided the component parts were kept clean and the carbide container emptied and cleaned out before being refilled. Electric lighting was, however, starting to catch on. It was easy to incorporate lighting coils in the flywheel magneto of a Villiers engine, to provide direct lighting. Alternatively, a Lucas Magdyno, a combined magneto and direct-current dynamo, driven by the engine, could be fitted to the larger-capacity machines.

Britain did not have a complete monopoly in motorcycle design. Many advanced designs originated from the Continent, amongst them the eight-valve twins made by Peugeot and Bleriot, the Alcyon two-stroke and the Swiss-made Motosacoche.

In Germany, interesting developments

An ambitious design that never went into production was this in-line four-cylinder overhead model made in 1922 by Vauxhall, the famous motor-car manufacturer. Of 930 cc capacity and with shaft final drive, the Vauxhall is alleged to have produced 30 brake horsepower and was capable of just over 80 mph (129 km/h). Only two complete machines were assembled, one of which has survived and been restored to running order.

One of the most colourful of all of motorcycling's personalities was Freddie Dixon, of Stockton-on-Tees. A brilliant engineer, an outstanding rider (and car driver) and possessed of great strength, Freddie's exploits after winning a race, both at home and abroad, are legendary. He won the first Sidecar TT riding a Douglas fitted with a banking sidecar of his own design. He also rode numerous other makes, which included this single-cylinder side-valve Indian on which he finished third in the 1923 Senior TT.

were taking place within the BMW factory in Munich. During the war BMW had made aircraft engines, which had earned praise from Baron von Richthofen, the German fighter ace. The Armistice terms required the company's diversification into other activities, which included making agricultural machinery and office furniture. They also tentatively entered the motorcycle market by making a 146 cc two-stroke engine. It made little impact, so they then designed and made the Helios, a complete motorcycle modelled closely on the contemporary Douglas horizontally-opposed twin. It led to greater things, and by 1923 they had available their memorable BMW R32 model, which put them on the road to success as a leading motorcycle manufacturer.

The United States could offer a challenge by way of their large-capacity vee-twins, until Henry Ford's Model T car appeared. The cheap, mass-produced motor car offered the American public a more satisfactory form of transport, and their motorcycle industry went into decline. Eventually the only major manufacturers to survive were Indian and Harley-Davidson.

Howard Davies made history when he won the 1921 Senior TT on a 350 cc AJS and rode the fastest lap in that year's Junior Race on the same machine. His luck deserted him in 1922, however, when he retired in both races. He is seen here on a 1922 AJS, the so-called 'big-port' model that everyone wanted. Note how the cylinder head is retained by means of a simple 'stirrup' arrangement, and the large-diameter exhaust pipe.

THE OVERHEAD-VALVE ENGINE BECOMES A REALITY

The overhead valve engine finally became a practical proposition in 1921, when AJS scored an overwhelming victory in the 1921 Junior TT Race. They took the first four places, and when they entered two of their Junior models in the Senior Race, they won that too.

If everything had gone according to plan, this result should have been achieved a year earlier, when their new overhead-valve engine made its debut in the 1920 Junior TT. Unfortunately fate intervened and all but eliminated the entire AJS entry, although Cyril Williams's engine did not pack up until he was 4 miles (6 km) from the finish on his last lap. He had amassed such a lead by then that by coasting and pushing he was still able to cross the line first. The novelty of the engine lay in having its overhead valves at an angle in the cylinder head, resulting in a combustion chamber of near perfect shape. It was claimed that these engines had a power output of 10 brake horsepower, high for a 350 at that time. It marked the early beginning of the legendary 'big-port' AJS, a giant-killer.

When AJS took first and second places in the 1922 Junior TT, few needed convincing that the overhead-valve engine was the way to go. Not only was it more powerful but also more efficient, using less petrol. Soon, overhead-valve models appeared in almost every manufacturer's range, although it would be a long time before its older side-valve counterpart was abandoned.

In the ever continuing search for greater efficiency, Triumph adopted a four-valve cylinder-head arrangement devised by Harry Ricardo, a leading authority on cylinder-head design. Never strong in racing, Triumph failed to achieve anything like the level of success of AJS, although one

Sunbeam also realised the potential of the overhead-valve engine and had one fitted into this Sprint Model for George Dance to ride at Brooklands. Dance was the leading sprint-racing competitor at this time, with innumerable 'fastest time of the day' victories to his credit. He was never quite so successful in road racing, and particularly unlucky in the Isle of Man.

Above: Six specially prepared BSA four-valve overhead-valve models started the 1921 Senior TT. They were unconventional in many respects, having knife-edge rocker arms with no provision for lubrication and a Y-shaped exhaust manifold that branched into two separate long exhaust pipes. None of the machines finished the race: all seized up. The BSA board banned further racing activities.

Kaye Don was another well-known racing man who also drove cars at Brooklands. Here he poses with one of the new lightweight 350 cc overhead-valve AJS models, similar to those that scored an overwhelming victory in the 1921 Junior TT. Their lightness in weight is obvious, resulting in a very favourable power to weight ratio that made them the 'giant-killers' of this era.

of their four-valve models finished second in the 1922 Senior TT. They had successes with them at Brooklands and other high-speed venues too.

Higher power output and greater speeds meant better engine cooling, with more heavily finned cylinder barrels and heads. Granville Bradshaw, who had originated the ABC, decided to go one better by designing an oil-cooled overhead-valve engine. It enjoyed a brief spell of popularity in the Dot and Alldays model ranges but was never completely accepted. Its oil content was too small, and it became known as the 'oil boiler'!

By 1923 it had become obvious that the Isle of Man TT had not kept to its original objectives. The machines now being entered by the factories were far removed from their standard production models and, furthermore, professional riders were being recruited to ride them. To give the amateur rider a better chance, it was decided to hold similar races later in the year, over the same $37\frac{3}{4}$ mile (60.75 km) Mountain Course, to be known as the Amateur TT.

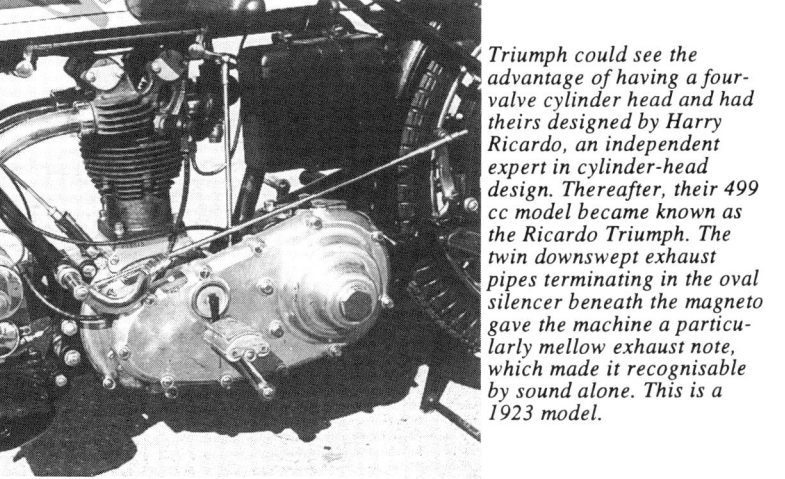

Triumph could see the advantage of having a four-valve cylinder head and had theirs designed by Harry Ricardo, an independent expert in cylinder-head design. Thereafter, their 499 cc model became known as the Ricardo Triumph. The twin downswept exhaust pipes terminating in the oval silencer beneath the magneto gave the machine a particularly mellow exhaust note, which made it recognisable by sound alone. This is a 1923 model.

There were companies that were not obsessed with racing and yet were still capable of making a good-quality, thoroughly reliable motorcycle. Sun is a particularly good example – a Birmingham-based company that had begun making motorcycles in 1906 and were well-known for their pedal cycles. The engines they fitted were bought in, and customers had a choice of a JAP or a Blackburne. This is a 1923 3$\frac{1}{2}$ horsepower model fitted with a side-valve Blackburne engine. The registration number has an oriental look about it!

Unquestionably one of the greatest riders ever seen, the Irishman Stanley Woods made his debut in the 1922 Junior TT, to finish a very creditable fifth on a Cotton. A year later, he won the Junior TT on a Blackburne-engined Cotton seen here. Like Alfred Scott, the small Cotton factory in Gloucester believed in using only straight tubes in the construction of its frames. Absolute rigidity gave a high standard of road holding.

Harry Reed of Manchester was one of motorcycling's pioneers, having founded Dot Motorcycles in that city during 1905. He had the right build to handle a racing sidecar outfit and entered the first Sidecar TT of 1923 on this Bradshaw-engined Dot attached to a Canoelet sidecar. He did well to finish fifth with an engine not renowned for its reliability.

It is difficult to imagine that at one time a member of the royal family entered a rider in race meetings at Brooklands. Here HRH the Duke of York (later King George VI) talks to his rider, S. E. Woods, before the start of a race at Brooklands in May 1922. The machine is a 994 cc eight-valve Trump-Anzani vee-twin.

There would be much stricter control over the category of rider accepted and the specification of the machine entered. It worked well for many years, until these regulations too became abused. By then its name had been changed to the Manx Grand Prix, which continues in use today.

Meanwhile, the TT programme underwent another change in 1923, the addition of a sidecar race run over the Mountain Course. Initially, it was greeted with enthusiasm, but after being run for three consecutive years it was abandoned. Sidecar manufacturers claimed to have received very little benefit from it and withdrew their support.

Brooklands now came into the spotlight. Used by both cars and motorcycles, not only for race meetings but also for long-distance record-breaking attempts, the track was often in use all through the night. Noise from unsilenced exhaust pipes gave rise to complaints from residents living close by, and the threat of legal action. A compro-

mise necessitated the obligatory fitting of a regulation silencer to every vehicle that used the track, the so-called 'Brooklands Can'. Specially designed to restrict exhaust noise, it did not affect engine performance to any marked extent.

The overhead-valve engine had given many manufacturers a new lease of life, not least Norton, who already had been scoring numerous racing and record-breaking successes with their side-valve models. In 1922 they produced their own overhead-valve engine, which proved so successful that it set Norton on the road to worldwide fame. For the next three decades Norton racing motorcycles were virtually unbeatable in international competition.

Many famous riders made their names during this era, including Stanley Woods, Freddie Dixon, Howard Davies, Alec Bennett, Graham Walker, Harry Langman, the Twemlow brothers and Bert Le Vack. Le Vack and Dixon were as good development engineers as they were riders, a rare

attribute. It was Le Vack who had revived the British vee-twin engine, making it outrun its American counterparts, when he was working for JAP.

Until 1925 it had been possible to close off short sections of the public highway for sprints and hill climbs, but an accident on Kop Hill in Buckinghamshire finally brought this to an end. Although no one was seriously injured, it highlighted the problem of crowd control and such competitions on public roads were stopped before a fatality occurred.

It seems strange that betting on race results was permitted at Brooklands, although in the early days race meetings there were run like horse races. 'Long Tom', a well-known book-maker at the track, is seen here with Kaye Don on his left and Ernie Bridgman on his right. The last named was closely associated with the Indian marque.

Freddie Dixon could capably handle a large-capacity vee-twin at Brooklands, despite its notoriously bumpy surface which necessitated a strong pair of wrists to keep the machine on a straight course at high speeds. Dixon prepared his own engines with meticulous care and appears to have designed and fitted an air box to the inlet system of his Harley-Davidson.

Although Alfred Scott's purpose-built gun car never received the approval of the War Office during wartime trials, he used its layout as the basis of the Scott Sociable three-wheel car. It was well made, as one would expect, and powered by a 578 cc twin-cylinder water-cooled two-stroke engine of his own design. Its decidedly odd appearance, like that of a car that had lost a wheel, proved a deterrent to sales.

EXPANDING OPTIONS

Two-stroke engine development had not kept pace with that of the four-strokes, largely because the foremost proponent of this type of engine, Alfred Scott, had died unexpectedly during August 1923, aged forty-eight, when an unattended cold developed into pneumonia. Since the war he had devoted most of his time to developing the Scott Sociable, based on the three-wheel gun car he had designed during the war. Although the Sociable incorporated many ingenious ideas, its appearance was more like that of a car with one wheel missing. It failed to attract the potential purchaser.

With so many different makes and models on the market, there was keen competition over price, which led to price cutting: Triumph took a lead by offering their new

When price-cutting in the motorcycle trade became rife in the mid 1920s, Triumph launched their 494 cc Model P in late 1924 with a price tag of £42 17s 6d. It represented good value for money, even if the early models like this had a front brake that was a contracting band of friction material around a pulley. Douglas tried to undercut this price, but Triumph responded with a further price reduction they could not match.

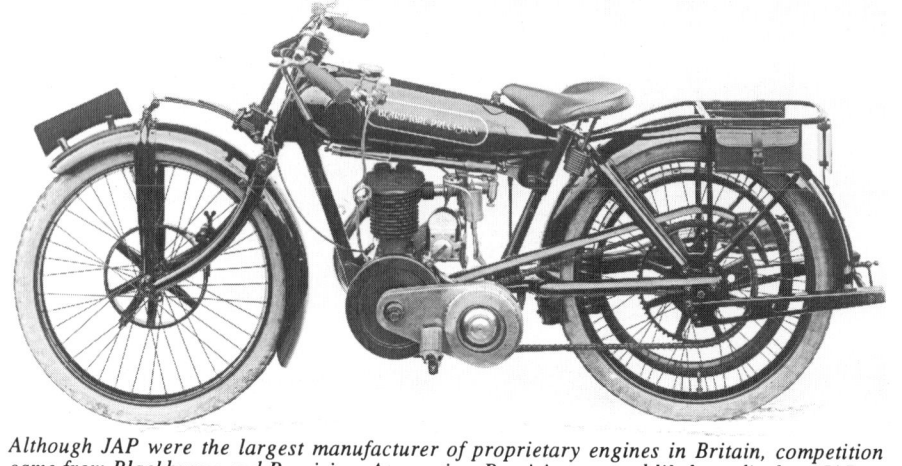

Although JAP were the largest manufacturer of proprietary engines in Britain, competition came from Blackburne and Precision. At one time Precision seemed likely to displace JAP as the market leader, but somehow the Tottenham-based factory retained their status. Precision also ventured into manufacturing complete motorcycles, such as this Beardmore-Precision of the mid 1920s. Was there ever an uglier motorcycle?

494 cc Model P for only £42 17s 6d. Many who had not done so before added a cheap basic-specification two-stroke to the lower end of their range. They could then claim they had something to suit everyone's taste (and pocket). Production had risen from 60,000 in 1922 to 110,000 in 1924 and by a further 10,000 to 120,000 in 1925, by which time there were 558,911 motorcycles and sidecar outfits in use.

The superiority of the overhead-valve four-stroke engine led towards further refinement, by having the valves operated by an overhead camshaft. It reduced the weight

of the valve gear and provided more accurate control of the valve timing. The idea was not new: Peugeot had raced double overhead-camshaft vertical twins in 1913, although their reliability had been questionable at that time. Others had tried too, but none so successfully as Percy Goodman, who patented his 348 cc design in 1925. It brought Veloce Ltd back into making four-strokes again, and after winning the 1926 Junior TT they had to move to larger premises to cope with all the orders. Two further wins in the Junior followed, in 1928 and 1929, to consolidate the busi-

When Veloce Ltd re-entered the four-stroke market with their new overhead-camshaft model in 1925, it bore the name Veloce on the petrol tank like its pre-1915 ancestors. No one recognised the name so the company had hurriedly to revert to 'Velocette' and capitalise on the success of the two-strokes on which they had built their reputation since their introduction in 1913.
This is an early flat-tank overhead-camshaft model with a Druid front fork.

Sunbeam designed an overhead-camshaft engine, a 500 cc unit for use in the 1925 Senior TT, and one of 600 cc capacity for that year's Sidecar TT. Designed by George Greenwood, they were not a success, largely as the result of unforeseen lubrication problems. They were known in the Sunbeam works as Crocodiles, as they made a tick-tock sound, like the watch swallowed by the crocodile in 'Peter Pan'.

They persevered with it for three years, but it was never a success and eventually was abandoned.

A further engine refinement at this time was the adoption of dry sump lubrication, so that the engine oil was circulated throughout an engine and then returned to the oil tank. Until then, total loss systems had been employed, in which oil was metered or pumped to an engine, then dispelled with the exhaust gases or discharged in small quantities on to the road.

JAP, who were second to none in providing engines of their own manufacture to the motorcycle trade, started to lose custom during 1924, when some of their larger customers began to design, and make, their own engines. Royal Enfield were one of the first to do so, with others following, so that eventually JAP were left with only the smaller customers who ordered in limited numbers. They had attempted to break into the two-stroke market during 1923 by offering their own two-stroke engines marketed under the name Aza, but without much success. Villiers were already too

ness. Incredibly, Alec Bennett had won the 1926 Race by a margin of more than ten minutes! The overhead-camshaft Velocette was years ahead of its time.

Harry Collier had designed a 348 cc overhead-camshaft engine for Matchless in late 1920. Its layout in the frame was unusual, the camshaft drive running up the rear of the engine, with the exhaust port on the nearside and the inlet port on the offside.

Abingdon King Dick began making motorcycles in 1903 and enjoyed a long association with the trade that continues today, albeit as manufacturers of workshop tools, such as spanners. On later models they used just their initials, AKD. Their big twin of 1922-3 was fitted with an unusual choice of engine, a side-valve Blumfield of 795 cc capacity.

Although Alfred Scott died in August 1923 from pneumonia after returning from a potholing expedition in damp clothes, the marque lived on. The Shipley factory entered a team in the Senior TT until 1931, but with no outstanding success other than a third place in 1927. Of the three riders seen here before the 1926 TT (left to right: H. Langman, E. Mainwaring and J. H. Welsby), only Welsby finished, in eighteenth place.

well established and had the reputation to go with it. Fortunately, much of JAP's business came from continental manufacturers, which continued, but within a decade they had to diversify into the manufacture of stationary engines to retain profitability. Surprisingly, they had also been making pencils on a vast scale since 1920. On the same site, Pencils Ltd were producing 1.5 million pencils a week during JAP's fiftieth anniversary in 1951.

To withstand the additional stresses caused by engines with greater power output and higher performance, the frame and front fork had to be redesigned. Frames constructed solely from straight tubes seem to give the least trouble and to offer a high standard of road holding, Cotton and Scott both being outstanding in this respect. Already, some models that had handled in an exemplary manner before were subject to speed wobbles at higher speeds, necessitating the use of a steering damper. The area requiring the most attention was the brakes, the old stirrup-type front brake mercifully having died out. Band brakes and those where a rubber pad pressed on to a dummy belt rim were not very efficient either, especially in the wet. The answer came in the internal expanding hub brake, which had been pioneered by H. C. Webb & Company, the fork manufacturers.

Whether or not alcohol fuels should continue to be allowed in the TT became the subject of much heated discussion throughout 1924 and 1925. Many thought it gave an unfair advantage and was not in accordance with the original rules of the TT. On the other hand, it was much kinder on engines, as they ran cooler and it allowed a much wider tolerance over mixture control. As will be seen, a firm decision was shelved until 1926.

The refinements were available to those who could afford them; for the rest, there was still good value to be had in some of the bargain-price models.

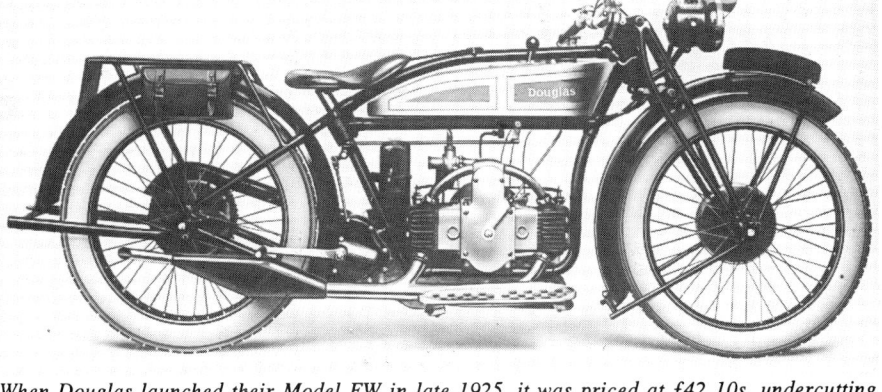

When Douglas launched their Model EW in late 1925, it was priced at £42 10s, undercutting Triumph's Model P by 7s 6d. Triumph responded by lowering their price by a further £1 7s 6d, which Douglas could not match. The EW sold in large numbers, but corners had been cut to achieve the low retail price. As a result, the early models were plagued by faults, which had to be put right under warranty. Even today, the EW has a bad reputation amongst those who remember it.

SADDLE TANKS AND WIRED-ON TYRES

Price cutting was rife by 1926. When Douglas brought out their new 348 cc Model EW in late 1925, they priced it at £42 10s, to undercut Triumph's bargain-price Model P by 7s 6d. Triumph responded by reducing their Model P's price by a further £1 7s 6d, to retain the competitive edge. The Triumph proved by far the better bargain, apart from being a single and having the larger capacity. Douglas had cut corners to get the price down, and their quality control was not all it should have been. The early models were plagued with faults, which Douglas had to rectify under warranty. The EW got a bad name, remembered to this day.

Although price cutting was very much the order of the day, it did not inhibit the development of new models. At the 1926 Motor Cycle Show, P & M, for example, created quite a stir with their unorthodox 246 cc vee-twin, its engine built in unit with a four-speed gearbox. Mounted transversely in the frame, the engine embodied a number of unusual features, not the least of which were leaf valve springs. A

Dunelt were another two-stroke manufacturer who went their own way, making their own engine with a truncated piston. It was claimed it provided a supercharging effect that gave the engine greatly increased torque. This is their 1925 499 cc Sports model, which had an aluminium alloy cylinder head and piston, and mechanical pump lubrication.

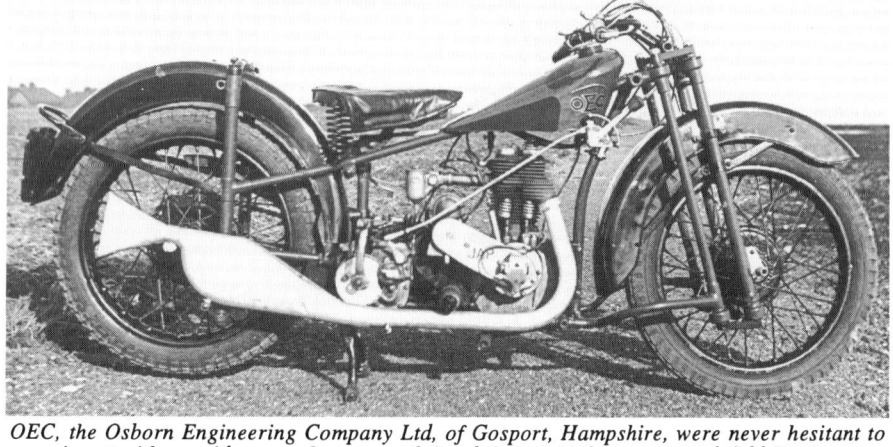

OEC, the Osborn Engineering Company Ltd, of Gosport, Hampshire, were never hesitant to experiment with new ideas, so that many claimed their initials represented Odd Engineering Contraptions! Their Duplex steering was one such innovation, a form of hub steering with provision for rear suspension. It had its advantages, but these were outweighed by its appearance and difficult handling at very low speeds.

Granville Bradshaw design, it broke with tradition in many ways but was a little too advanced to appeal to most motorcyclists, always wary of change.

OEC were experimenting with duplex steering, a form of hub steering which Wallis and others had also tested. The first model was on view at the 1927 Show, but there were not many takers. Its virtues were overshadowed by its appearance and lack of response at low speeds.

Changes in design were now beginning to influence the appearance of the motorcycle. When Howard Davies began making the JAP-engined HRD models that bore his initials in 1924, he had fitted them with a saddle tank, an improvement as it no longer had to be squeezed between the top tubes of the frame. George Brough had helped his father to make motorcycles bearing the family surname, but he left to start making his own in 1919, using the some-

After his TT successes with AJS, Howard Davies set up his own company to manufacture HRD motorcycles, which bore his initials. He could not have achieved a better start, as in his first full year of production he won the 1925 Senior TT on this JAP-engined model. He was not in business for long, succumbing to the depression of the late 1920s. The company was bought eventually by Philip Vincent through an intermediary.

The 1925 Sidecar TT proved to be the last for many years as the trade had shown little interest and it did not help sell more sidecars. No doubt the acrobatics performed by the sidecar passenger would have proved a deterrent! The driver seen here at Ramsey Hairpin is Jimmy Simpson, who remained faithful to AJS. He finished fifth, which must have pleased him, as he was hard on the machines he rode and often retired as the result of engine failure.

what unkind title 'Brough Superior'. He too had used saddle tanks from the start, of a unique shape that became his trademark, although Coventry-Eagle subsequently made and fitted a passable copy to their own vee-twins. A saddle tank lent itself to greater freedom of styling as its curvaceous lines could be decorated more attractively. More to the point, it had a greater petrol-carrying capacity. The idea caught on and by the time of the 1928 Show the saddle tank had become the rule rather than the exception.

Another change that came into effect during 1927 was the use of wired-on tyres, to replace the beaded-edge type that had been in use for so long. Although beaded-edge tyres were easy to fit, they would come off a rim just as easily, and often quite unexpectedly. Only by pumping them up hard, and fitting three security bolts to each wheel, could they be retained reasonably well on the wheel rims of the more sporting models.

A wired-on tyre required a wheel rim with a different cross-section that had a well in its centre. More difficult to fit until the knack had been acquired, a wired-on tyre stayed put once fitted. It could be run safely at lower pressures and presented a larger section in contact with the road. Such were its advantages that within a year it had become the only kind of tyre fitted on all new motorcycles.

The overhead-camshaft engine had shown its full potential in 1926, when Dougal Marchant became the first person to lap Brooklands at over 100 mph (160 km/h) on a 350. He had done so on a Chater-Lea, fitted with an engine he had developed himself. In 1927 Norton had their own overhead-camshaft engines available for the Senior TT, designed by Walter Moore, who had earlier associations with Douglas and ABC. AJS had their design available too, in time for the same race, having the camshaft driven by chain rather than by a vertical shaft. Only Sunbeam appeared to have had little luck with this type of engine and had abandoned it after all their works-sponsored entries had failed in the 1925 TT. The Velocette design was still far superior to all the others in the 350 cc capacity class and gained a further advantage

Dougal Marchant became the first rider to exceed 100 mph (160 km/h) on a machine fitted with a 350 cc engine. In 1926 he used a much modified Blackburne engine fitted into a frame and cycle parts made by Chater-Lea. Here he is seen at Brooklands, where he won the 350 cc Solo Championship in September 1926 at 93.97 mph (151.23 km/h). His best flying lap speed was 96.90 mph (155.94 km/h).

The 490 cc side-valve Norton, with its direct belt drive to the rear wheel, had proved surprisingly fast for such an elderly design. However, the advent of the over-head-valve engine brought to a close its long run of successes. Norman Black had the honour of riding the last belt-drive model to finish in the TT, as far back as 1920. Compare with the following photograph.

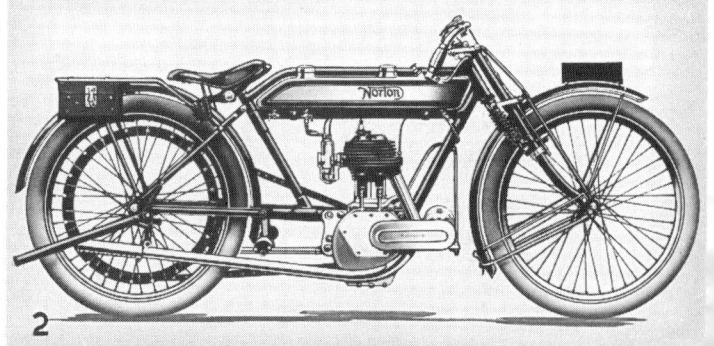

An experimental overhead-valve Norton was made in 1922 and sent to Brooklands for test. Initial reaction was cool, but by 1923 the Norton works team were all equipped with overhead-valve models for that year's senior TT. By 1925 the new model had many successes to its credit and had become a force with which to be reckoned. Jimmy Shaw rode this machine into second place in the 600 cc race of the 1925 Ulster Grand Prix, its diminutive front brake seemingly no disadvantage.

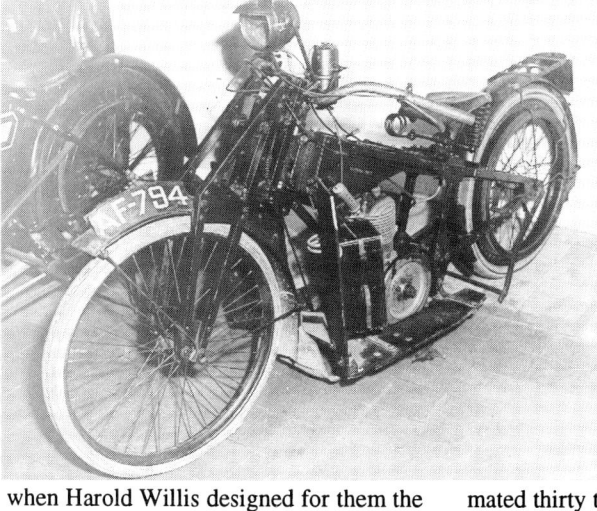

If you cannot afford a new motorcycle, make your own! W. J. Battershill, of Yelverton, Devon, did just that in 1926. He fitted a 1917 2¹/₄ horsepower Union single-cylinder two-stroke engine into an ashwood frame and fork assembly strengthened with gusset plates, which he had made himself. Named the Yelverton Union, it has survived, to become one of motorcycling's curiosities.

when Harold Willis designed for them the first practical positive stop foot gearchange.

The event that had the most impact in 1928 was the inaugural speedway meeting held on 19th February at High Beach, in Epping Forest. Although it was not the first speedway meeting to be held in Britain, it was the first to make any lasting impact and draw a vast crowd. The organising Ilford Motor Cycle and Light Car Club had expected less than two thousand spectators, but to their amazement an esti-mated thirty thousand came! It was not an outstanding success, as no one yet knew how to prepare a suitable track, but the daredevil exploits of the riders captured the imagination of the public. Before the year was out, tracks were being made in every major city and town. This new sport, which had been brought over from Australia, was in Britain to stay, with attendances second only to football. Few other motorcycling events have made the front page of the national daily newspapers!

A new sport, dirt-track racing, swept Britain in 1928, drawing huge crowds of spectators. Tracks opened up in almost every major town and city. The most successful machine was the purpose-built Dirt Track Douglas, a 494 cc horizontally-opposed twin that lent itself ideally to the leg-trailing riders who rode on a bed of cinders. These machines, which were broadsided on the corners, ran on methanol fuel and had no brakes. The rider is Ron Hieatt, his name engraved on the air-intake box to the carburettors.

BSA's 249 cc single-cylinder side-valve model launched in 1925 proved an enormous success. It sold for £36 10s and was aimed at those to whom price was a major consideration. Countless newcomers to motorcycling learnt to ride on one of these round-tank models, which helped to establish an early brand loyalty to the BSA marque.

THE LAST YEARS OF THE VINTAGE PERIOD

The years 1927 to 1930 were the golden age of vintage motorcycling. During that four-year period the motorcycle had developed into a cheap, reliable, economic and ubiquitous means of transport, with a camaraderie amongst riders that today seems almost unbelievable. From 1927 to 1929 the number of motorcycles in use had risen from 681,410 to 731,298, a figure that was not surpassed until 1950. Although the number fell to 724,319 in 1930, it fell much more sharply in the years that followed, when the recession was at its worst, accompanied by a high level of unemployment.

Some well-known makers had already gone out of business, whilst others, AJS and Humber amongst them, would soon follow. Even Sunbeam were in financial difficulties and had been bought by ICI in 1928. But exports of British motorcycles had risen to a record peak in 1930, before they too took a sharp downward turn.

After a long period as the industry's leading manufacturer, Triumph found themselves in decline and displaced by BSA. BSA had achieved record sales in 1925 with their 'round-tank' model, which was cheap and sold in large numbers. Many

people began their motorcycling apprenticeship on one, and from this a brand loyalty was established. When BSA brought out their 'sloper' models in the late 1920s, these too offered good value for money and established a trend – that of an engine inclined at a forward angle in the frame. Most of the other leading manufacturers soon followed suit.

Rudge were now in the ascendancy with their four-valve engines, a highlight being Graham Walker's win in the 500 cc class of the 1928 Ulster Grand Prix at over 80 mph (129 km/h). Ariel had some very good models too, their design team being headed by Val Page and including Edward Turner and Bert Hopwood. Royal Enfield had a comprehensive range of models, from their 225 cc two-stroke to a massive 976 cc vee-twin, whilst new Scott Super Squirrel twins were available with either a 496 cc or a 596 cc engine and a three-speed gearbox.

In 1929 Veloce Ltd surprisingly put on general sale a road-racing replica, the KTT, virtually identical to the models raced by the factory. It encouraged many amateur riders to take up racing comparatively cheaply, with more than a fair chance of

Rudge, more than any other firm, made the most of the four-valve head arrangement, as seen here in their 1928 500 cc model. Graham Walker won the 1929 Ulster Grand Prix on a 500 cc four-valve Rudge, which became the basis of the famous 500 cc Rudge Ulster model. Rudge also took the first two places in the 1930 Senior TT, the last occasion on which a Senior TT race was won by a machine with pushrod-operated valves.

success. Douglas produced a model that was just as successful on the speedway, their famous Dirt Track model that was ideally suited to the leg-trailers of that era. It dominated racing on the cinders for the next couple of years and 1300 had been sold by the end of 1929.

Howard Davies had sold his interest in HRD motorcycles to Ernie Humphries of OK Supreme, to be bought by Philip Vincent. Vincent foresaw the advantages of a spring frame, and it became a strong selling point in the HRD's favour. George Brough was busy originating new designs

Velocette's memorable contribution in 1929 was to offer the first 'over the counter' production racer to be made in any quantity. Their overhead-camshaft 348 cc KTT model was an outstanding success, having the first foot-operated positive stop gearchange. The design of the KTT can be attributed to Percy Goodman, being closely based on his original overhead camshaft 'K' series design of 1925.

George Brough started by helping his father make the Brough motorcycle, a horizontally opposed twin of 496 cc capacity. Convinced he could do better, George started to make his own motorcycle in late 1919 and called it the Brough Superior. Built largely from bought-in parts, its build was of a high standard and it became known as the Rolls-Royce of motorcycles. It was a large-capacity vee-twin, fitted with a JAP engine. George Brough is on the right of record-breaker Joe Wright.

that broke new ground. A transverse-mounted vee-four was soon followed by an air-cooled side-valve in-line four, with a spring frame, intended primarily for hauling a sidecar. His more adventurous models were made alongside his vee-twins and were priced accordingly. It was no coincidence that the Brough Superior was known as the Rolls-Royce of motorcycles, and officially endorsed as such by Rolls-Royce themselves!

One of the most ingenious of all the 1929 models was the Ascot Pullin. The creation of Cyril Pullin, it was ultimately unsuccessful because it was considered too advanced for its time. It comprised a 496 cc horizontally mounted overhead-valve engine in unit with a three-speed gearbox,

enclosed within a pressed steel frame, with a pressed steel front fork and handlebars. Both wheels had hydraulically operated brakes and were fully interchangeable, the final drive chain also being fully enclosed within the frame pressing. At £75 it was not cheap.

The weight limit was raised from 200 pounds (90 kg) to 224 pounds (102 kg) for machines eligible for the £1 10s per annum road tax rate, but this was soon repealed after a change of government. Far more widely appreciated was the abolition of the 20 mph (32 km/h) speed limit, long overdue.

In 1930 chromium plating was introduced, far more durable than the nickel plating it replaced. Its brightness was so

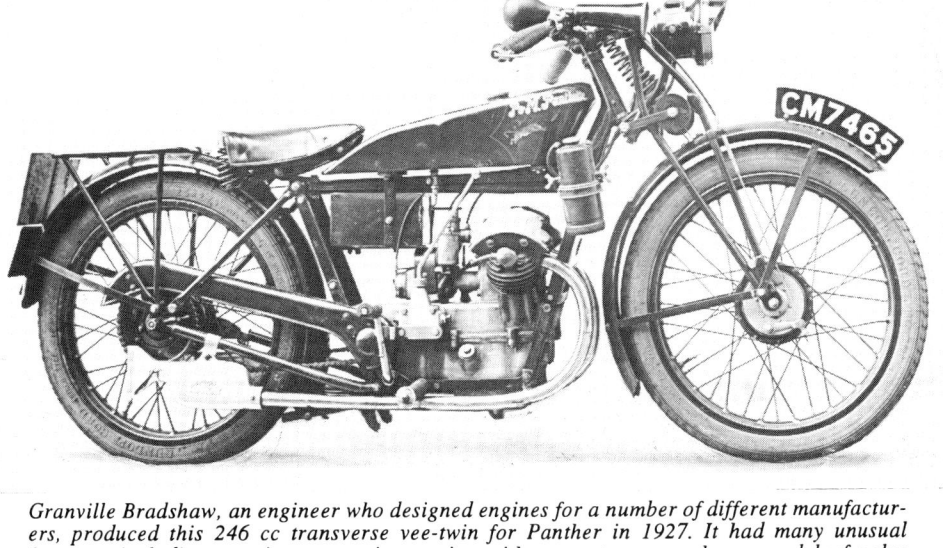

Granville Bradshaw, an engineer who designed engines for a number of different manufacturers, produced this 246 cc transverse vee-twin for Panther in 1927. It had many unusual features, including a unit-construction engine with a car-type gearchange and leaf valve springs. Silent-running, it attracted much attention but not enough buyers to make it viable. By late 1928 its frame was housing instead a 147 cc Villiers engine.

Matchless sprang a surprise at the 1929 Motor Cycle Show when they unveiled a new spring-frame model that looked at first glance like a vertical twin. Their Silver Arrow was actually a narrow-angle (26°) side-valve vee-twin. Although quiet-running, its 400 cc engine provided a disappointing performance and it was never a very popular model. To all intents and purposes it was dropped from production in 1933.

The main attraction at the 1930 Motor Cycle Show was the 499 cc Ariel Square Four designed by Edward Turner. Its overhead-camshaft engine had its cylinders arranged in block formation, like two vertical twins back to back in a monobloc casting. Like so many promising designs, the production version differed in many respects from Turner's prototype, to effect economies. Even so, it was an outstanding model that continued in various forms as Ariel's 'flagship' over the years. This is a 1931 model.

striking in appearance that the more chrome a motorcycle had the more likely it was to sell! If a twin-port cylinder head could be fitted to an engine, this had the added advantage of needing two exhaust systems, one on either side. It was all very irrational, but if that was what the public wanted and it helped sell motorcycles in difficult times, what did it matter?

Matchless produced a unique spring-frame design in 1930, their 400 cc Silver Arrow model. Outwardly it looked like an in-line twin, not the narrow-angle side-valve vee-twin it actually was. Quiet-running, the Silver Arrow created quite a sensation, but in practice its performance was hardly sparkling. Only 1400 were sold that year.

Also of considerable interest was the 500 cc Ariel Square Four, an overhead-camshaft design originated by Edward Turner, of whom

a great deal more would be heard in future years. It had its cylinders arranged in the form of a cube, two at the front and two at the rear, in a massive monobloc casting. It was destined to continue in production for almost three decades, increasing in capacity to 997 cc and reverting to pushrod valve operation before production ceased.

Two manufacturers, New Hudson and Triumph, opted for partial enclosure, having the lower half of the engine, and the gearbox, fully enclosed by removable metal covers. Not only was the machine easier to clean, by hosing it down, but it also saved the manufacturers having to polish the timing and gearbox end covers. The idea never caught on, although Velocette tried it some twenty-five years later, with equal lack of success.

Velocette introduced their first two-stroke with its crankshaft supported at both ends and the novelty of coil ignition, the

GTP. They also pandered to public taste with a twin exhaust-port version of their overhead-camshaft model, the KTP, also having coil ignition. The latter was one of their rare failures, discontinued a year later in 1931.

The world's absolute speed record for motorcycles rarely received much attention until a successful attempt had taken place. At the beginning of 1930 Ernst Henne held the record on a supercharged BMW at 134.68 mph (216.74 km/h). Determined to regain it for British prestige, Joe Wright raised it by 2.64 mph (4.25 km/h) on an OEC-JAP, only to have Henne cap it by a mere 0.34 mph (0.55 km/h) shortly afterwards. Refusing to give in, Wright raised it higher still, to 150.74 mph (242.59 km/h), in time for the record-breaker to be displayed on the OEC stand at that year's Motor Cycle Show. Yet it was not the OEC-JAP that had broken the record! Wright had taken two machines with him and was riding the reserve one, a Zenith-JAP, when he recorded the record-breaking speed. The reason for this deception was never revealed and is still the subject of controversy.

Regular competitors at Brooklands faced a setback at the end of 1930 when the machine manufacturers and accessory suppliers announced that they would no longer pay retainers or make bonus payments for racing successes. Some riders had already deserted Brooklands when the use of alcohol-based fuels had been banned in the TT, as the same engine could no longer be used at both venues. Retainers and bonus money were an essential part of a rider's livelihood, so even more of them either retired

Francis Barnett came out with a novel idea in the late 1920s by having a frame comprised of a series of tubes bolted together. If the frame was damaged in an accident, it was easy to effect a repair merely by replacing the damaged tubes. The machine shown here is the rare 344 cc model of 1928 that used a semi-unit construction in-line twin-cylinder two-stroke engine specially made by Villiers. Not many were sold and in consequence few have survived.

29

Sunbeam were losing their status in the industry by the late 1920s as their designs began to stagnate. They had been acquired by ICI in 1928, but it was not until 1931 that any real effect began to be felt. Charlie Dodson scored their last Senior TT wins in 1928 and 1929; he is shown here after winning the 1928 race, with (left) designer George Greenwood and (right) Dick Birch, who had ridden for Sunbeam in 1927 and helped them win the Manufacturer's Team Prize.

or no longer used Brooklands as their base.

With commendable foresight, the Sunbeam Motor Cycle Club held the first of their Pioneer Runs in 1930, from Croydon Airport to the outskirts of Brighton. The event was run in conjunction with the Association of Pioneer Motor Cyclists, which had been formed at the 1928 Motor Cycle Show. It proved to be the first significant move towards the preservation of old

motorcycles and is still run today, as the opening meeting of the year.

Today, a machine can be classified as a vintage machine only if it has been manufactured between 1st January 1915 and 31st December 1930. These demarcation lines have stood the test of time well and seem unlikely to be changed in the future. Additional classes have been created to cater for machines of later manufacture.

FURTHER READING AND CLUBS

There are no books currently in print that relate exclusively to vintage motorcycles. However, a few books about old motorcycles have been written on a broader aspect, almost all in the past and no longer in print, and have become collectors' items. As they can sometimes be found for sale at autojumbles, motorcycle exhibitions and book fairs, details are given here:

Clew, Jeff. *The Restoration of Vintage and Thoroughbred Motorcycles.* Haynes, 1976.
Hough, Richard, and Setright, L.J.K. *A History of the World's Motorcycles.* Allen & Unwin,1966.
'Ixion'. *Motor Cycle Cavalcade.* Iliffe, 1950.
Setright, L.J.K. *Twistgrip.* Allen & Unwin, 1969.
Sheldon, James. *Veteran and Vintage Motor Cycles.* Batsford, 1961.
Various. *The History of Motorcycling.* Orbis,1979.

There are also a number of marque histories about individual makes of motorcycle, available from publishers such as Aston, Crowood, Haynes and Osprey.

MAGAZINES
There is no publication that deals specifically with vintage motorcycles. The following titles do, however, often carry articles on these machines (as does *The Vintage Motor Cycle*, which is available only to *bona fide* members of the Vintage Motor Cycle club):

British Bike Magazine, PO Box 19, Cowbridge, South Glamorgan CF7 7YD.
(The) Classic Motor Cycle, Bushfield House, Orton Centre, Peterborough, Cambridgeshire PE2 5UW.
Classic Motorcycling Legends, 80 Kingsway East, Dundee, Angus DD4 8SL.
Old Bike, Clayside Barn, Alstonefield, near Ashbourne, Derbyshire DE6 0AA.
Old Bike Mart, PO Box 99, Horncastle, Lincolnshire LN9 6LZ.

CLUBS
The Association of Pioneer Motor Cyclists, Heatherbank, May Close, Liphook Road, Headley, Bordon, Hampshire GU35 8LR.
The Sunbeam Motor Cycle Club, 18 Chieveley Drive, Tunbridge Wells, Kent TN2 5HQ.
The Vintage Motor Cycle Club Ltd, Allen House, Wetmore Road, Burton-upon-Trent, Staffordshire DE14 1SN. (Interest extends to all motorcycles more than twenty-five years old.)

There are also a number of local British motorcycle clubs and one-make clubs that relate to a specific marque or to a particular country of origin.

PLACES TO VISIT

Intending visitors are advised to ascertain the times of opening and whether items of particular interest will be on display before making a special journey.

Birmingham Museum of Science and Industry, Newhall Street, Birmingham, West Midlands B3 1RZ. Telephone: 0121-235 1661.

Haynes Motor Museum, Sparkford, near Yeovil, Somerset BA22 7LH. Telephone: 01963 440804.

National Motorcycle Museum, Coventry Road, Bickenhill, Solihull, West Midlands B92 0EJ. Telephone: 01675 443311.

National Motor Museum, John Montagu Building, Beaulieu, Brockenhurst, Hampshire SO42 7ZN. Telephone: 01590 612345.

The Sammy Miller Museum, Gore Road, New Milton, Hampshire BH25 6TF. Telephone: 01425 619696.

Science Museum, Exhibition Road, South Kensington, London SW7 2DD. Telephone: 0171-938 8000.

Science Museum, Red Barn Gate, Wroughton, near Swindon, Wiltshire SN4 9NS. Telephone: 01793 814466.

Stanford Hall Motorcycle Museum, Stanford Hall, Lutterworth, Leicestershire LE17 6DH. Telephone: 01788 860250.

A group of motorcyclists pose in the Dorset village of Beaminster during the mid 1920s, probably at the start of a road trial as all are carrying riding numbers. Cloth caps are obviously popular wear, usually worn with the peak at the back and retained on the rider's head by goggles. The liveried chauffeur with his peaked cap looks a little out of place.